Pankaj Kadam

Human Machine Interaction Using Hybrid Biological Signals

Next step of powered wheelchair controller

Anchor Compact

Kadam, Pankaj: Human Machine Interaction Using Hybrid Biological Signals: Next step of powered wheelchair controller. Hamburg, Anchor Academic Publishing 2013
Original title of the thesis: Powered Wheelchair Controller Using Hybrid Bio-Signals

Buch-ISBN: 978-3-95489-143-6
PDF-eBook-ISBN: 978-3-95489-643-1
Druck/Herstellung: Anchor Academic Publishing, Hamburg, 2013
Additionally: School of Computer Science and Electronic Engineering, University of Essex, England

Bibliografische Information der Deutschen Nationalbibliothek:
Die Deutsche Nationalbibliothek verzeichnet diese Publikation in der Deutschen Nationalbibliografie; detaillierte bibliografische Daten sind im Internet über http://dnb.d-nb.de abrufbar.

Bibliographical Information of the German National Library:
The German National Library lists this publication in the German National Bibliography. Detailed bibliographic data can be found at: http://dnb.d-nb.de

© Anchor Academic Publishing, ein Imprint der Diplomica® Verlag GmbH
http://www.diplom.de, Hamburg 2013
Printed in Germany

Acknowledgement

In my endeavour to study the fundamental intricacies of this project and at various stages of implementation, I was offered great support and help. I am grateful to all and wish to thank them for their part in the success of this project.

I would like to thank Dr. Huosheng Hu, my supervisor, for his support, inspiration and rendering confidence in me.

Mr. Robin Dowling, Robotic Arena technician, for his technical support and assistance.

My parents and sisters, for their inseparable support and encouragement.

Last but not the least, all my friends who motivated me to carry on, at difficult times.

Pankaj Kadam

Abstract

The idea of using a powered wheelchair, for people with mobility limitation and the elderly has been around for quite a while. Most of these wheelchairs require the use of upper limbs strength to control them. The mobility needs of quadriplegic individuals are fully dependent on their care givers. This project aims to help quadriplegic individuals use their wheelchair with minimum human assistance and provide them a sense of mobility independence.

The project involves the use of the natural biological signals (Bio-signal) to control a powered wheelchair. The wheelchair is manoeuvred using brainwaves, eyes and facial muscles movements.

Bio-signals are obtained from the forehead using a special headband with electrodes. The headband provides an alternate communication channel between the operating person and the wheelchair. The signals obtained are filtered, processed and fed to an artificial neural network algorithm to generate control commands for the wheelchair. The operator is able to drive forward, reverse, left, right and stop the wheelchair. The main challenge involves real-time extraction and execution of the commands to control the wheelchair.

Keywords: Human Machine Interaction, Intelligent wheelchair, EEG, EMG and EOG.

Table of Contents

Table of Figures

Nomenclature

HMI: Human Machine Interaction

HCI: Human Computer Interface

BCI: Brain Computer Interface

EEG: Electroencephalography, measure neural activity of the brain

EMG: Electromyography, measures muscle activity

EOG: Electrooculography, measures resting potential of the retina

ANN: Artificial Neural Network

FANN: Fast Artificial Neural Network

CMRR: Common Mode Rejection Ratio

SNR: Signal to noise ratio

MIPS: Microprocessor without Interlocked Pipeline Stages

DSP: Digital Signal Processor

ADC: Analog-to-Digital Convertor

VoIP: Voice over IP, Internet data service for making telephone calls

WBAN: Wireless Body Area Network

1   Introduction

The current surge and advancement of technology has increased the social demands for the quality of life. This has given rise to the development of consumer conscious gadgets for everyday use, such as the latest mobile phones. These devices have made our life easier, faster, safer and more entertaining with improved user experience. As part of the efforts to improve the quality of life for the disabled and the elderly, robotic researchers have been trying to merge the robotic techniques into systems that can assist them in their daily life. The latest developments in research areas such as computer science, robotics and Artificial Intelligence have broadened the possibility to support disabled and elderly people with new assistance systems [1].

The assistance robotic system has to be safe and reliable to use but also user friendly. For this the user should have certain degree of control over the system to overwrite undesired actions of the machine. This is achieved by providing a prototype model of the system and giving it to the actual user for testing for a period of time and improvements in the design are made from their feedback. These types of systems are called the human-in-the-loop control system. These systems need to be tested thoroughly before any commercial production to meet standard requirements [2]. Human Machine interaction (HMI) is fast becoming one of the prominent technologies used for improvising the available resources.

The keyboard and mouse are often used as the Human Computer Interface (HCI) devices. However, it requires more training for the disabled and the elderly to get familiar with a computer. With the improvement of the processing power of computers, many researchers have tried to use computer vision, voice recognition and similar techniques. However the techniques have some flaws. Voice recognition systems are slow in interpreting the results and are easily affected by noise e.g. during a party. The vision based HCI still has to overcome the detection of the individual in real world environments with changing light conditions. Other researchers have proposed to use Bio-signals such as Electromyogram (EMG) [3], Electroencephalogram (EEG) [4], and Electrooculograph (EOG) [5] for HCI. Each of the Bio-signal has its own uniqueness which is used for extracting eminent information. Current researchers are trying to tap these by improving their detection and classification methods with the aid of new technology.

In the Bio-signals used for this project, EEG refers to the recording of the brain's electrical activity measured using multiple electrodes placed on the scalp region. The brain is always active and generates signals of different intensities. Researchers' are now able to identify different regions of the brain which are activated when we use different sensory organs or think of using them. EEG signal is produced by the triggering of neurons within the brain. They are widely used for diagnostic of epilepsy patients'. For better usability of the signals, we have to take into account the quality of the EEG recorded, user involvement and more accurate ways of signal analysis. EOG signals are mostly obtained by using two electrodes which are placed on the forehead region. When the eyes move towards one of the sides, it gets closer to one of electrode and away from the other electrode thus creating a potential difference. The signals are then compared with the resting potential of the retina and the movement of the eye in a particular direction is detected.

EMG refers to the recording of the electrical activity of the skeletal muscles, when they are electrically or neurologically activated. These signals are used for applications involving understanding of proper motion of muscle group, controlling bionic parts for the amputees, etc. These signals are strong compared to EEG and easy to notice. EMG signal are often applied to the rehabilitation system, e.g. electric prosthetic hand, because it can be generated by voluntary muscle contraction and it has better properties such as, high amplitude and signal to noise ratio (SNR) than other Bio-signals. EMG signals can be efficiency used as control command with higher accuracy has been concluded in previous research papers [3], [6] and [8].

In this project EMG, EOG and EEG signals generated by eyes and facial muscle movements are recorded using the head band with embedded electrodes. For a paralyzed individual the forehead is the most crucial area to capture useful signals. Also the headband appears less evident when used in public places compared to the electrode cap used in Brain Computer Interface (BCI). The signals recorded are further processed to extract unique features that can be used to control the wheelchair and are easier to be replicated by quadriplegic individuals. These features are given to the ANN to obtain the relevant decision logic. The logical table is used to maps particular movements with the control commands of the wheelchair. The following sub-parts explain more about different wheelchairs, their limitations and current research. The later chapter will give more detail about the controller algorithm software, hardware, GUI, testing of the entire system and future developments.

1.1 History

The first record of combining wheels to furniture was a Greek vase image of wheeled child's bed around 530 B.C. A picture of the painting is shown in figure 1.The use of wheelchair by people with activity limitation mainly started in the early 1900s. Since then manual wheelchairs have undergone many changes to fit the need of today's user. Despite disability, wheelchair helps persons to maintain mobility and have a social life. From the manual wheelchair, we have now moved to electric powered wheelchair. The needs of many disabled individuals are satisfied by use of manual or powered wheelchair, but there is a segment of disable individuals who cannot use them for independent mobility. To help these individuals, researchers are using technologies applied in the fields of mobile robots to build intelligent wheelchair with embedded devices and sensors. Nowadays even the traditional wheelchairs are available with improvements such as light weight structure, postural stability, efficiency in propulsion and portability in cars [9] [23].

Figure 1: Child's bed on rollers. From a hydria, Ionian made 530B.C [9].

1.2 Types of wheelchairs

1.2.1 Manual wheelchairs

Manual wheelchairs are propelled by the occupant by turning the hand-rims or by an attendant using handles. These are mostly used by individuals with upper-body mobility. They are of two main subtypes, rigid and foldable. Rigid chairs are preferred by active users as they have less moving parts and are light in weight. Foldable chairs are easy for storage or placement into a vehicle during travel and are mostly used at airports and hospitals. These wheelchairs can also be fitted with shock absorbers, to cushion bumps on the path [10]. The light weight wheelchairs reduce shoulder and wrist injuries due to strain, decrease the total energy expenditure and are easier for transportation.

1.2.2 Electric-powered wheelchairs

An electric-powered wheelchair uses an electric motor, a handle bar or joystick for navigation and is powered by batteries. Electric wheelchairs are used to travel a longer distance without physical exhaustion. These wheelchairs are useful for individuals who are too weak or otherwise unable to move around in a manual wheelchair. These wheelchairs are also provided to persons with cardiovascular conditions.

Furthermore Electric wheelchairs can be customized to cater individual needs by adding suspension to the front and back wheels, cushioning, light weight frame, pneumatic tires for softer rolling resistance, etc. There are also sports varieties built for wheelchair athletics, playing tennis, basketball, etc [8].

1.2.3 Limitation of the electric chairs

- Steered in an upright posture, hand strength and upper-body mobility are required.
- Mobility scooters have longer length, which limits their turning radius in smaller lanes.
- It has a low ground clearance that can make it difficult to navigate around poor structured paths.
- They have fewer options for body support, such as head or leg rests.
- They are quite heavy and not portable.
- Need to be charged regularly.

1.2.4 Smart or Intelligent wheelchair

A Smart wheelchair is a motorized chair with an superficial control system designed to assist the user. This control is generated with the help of software running on special computer accompanied with sensors and applying technology from the fields of robotics. The user can interact with the system using a joystick, touch sensitive display, a sip-and-puff device, etc. For obstacle detection and avoidance sonar, infrared sensors or Lasers are implemented. Some wheelchairs are attached with robotic manipulators, usually a robotic arm to grab household things.

Smart wheelchairs are designed with specific user requirements in mind. For a user with cerebral palsy the role of the smart wheelchair is to interrupt small muscular signals as high-level commands and execute them. Different techniques can also be implemented on a smart wheelchair such as face detection, path finder, artificial reasoning or behaviour based control techniques [8]

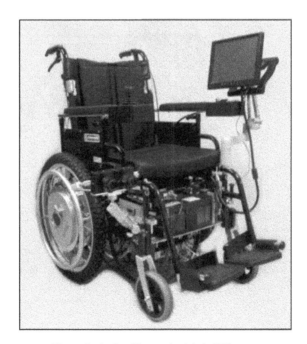

Figure 2: An Intelligent wheelchair [27].

Earlier intelligent wheelchairs were developed by adding a seating arrangement to mobile robots. The VAHM wheelchair belongs in this group. Its earlier models consisted of a wheelchair with a mobile robot base. It had three control modes, manual mode, semi-automatic mode and full automatic mode. In full automatic mode, autonomous navigation is based on internal maps. Semi-automatic mode involves wall following and obstacle avoidance [24].

The later models of smart wheelchairs were modified commercial powered wheelchair, with added functionality. The figure below shows a powered wheelchair with robotic arm which can grasp a bottle and pickup books and other similar objects.

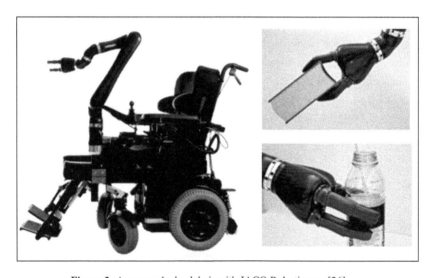

Figure 3: A powered wheelchair with JACO Robotic arm [26].

The other variety available is called the "add-on" unit. These units can be easily assembled or detached to any commercial wheelchair. These types of units are valuable for children. As they continue to grow, their wheelchair needs to be altered according to their needs. The figure below shows the Smart Powered Assistance Module (SPAM) for manual Wheelchair [23].

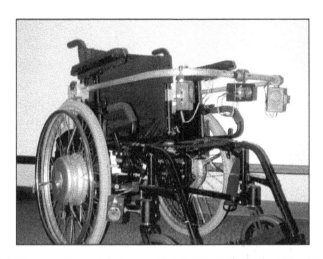

Figure 4: The Smart Powered Assistance Module (SPAM) for manual Wheelchair [28].

1.3 Current research

An intelligent software algorithm to control a wheelchair is indeed a challenging and significant problem, which has triggered attention of the research community. The current researchers are working on two arenas, first the control system for better utilization of the recorded Bio-signals such as feature extraction, classification algorithm, etc. The other area of research interest is the user-interface, adding different ways a user interacts with the machine such as hand gesture detection, voice recognition, etc. Solving safety concerns and usability issues without major changes in the user's environment are also important areas under development.

The simplest method used earlier was to activate a switch when certain electrical activity goes above a threshold. For better control more sophisticated methods are applied. Researchers are still tackling problems of real world implementation. Some of the problems faced are cross talk between the nearby electrodes, random variation in the muscle activation, noise due to muscle movements, change in the skin resistance due to sweat and many more. Some of the methods used to overcome these are, the use of differential electrodes, finding a suitable location for electrode placement, etc. Many of the other researchers have implemented different methods such as the NavChair project which uses Vector Field Histogram (VFH) for obstacle avoidance and navigation assistance for powered wheelchair. It uses obstacle avoidance algorithm developed for autonomous robots

[25]. This method finds a trade-off between the best path and the user's goal with obstacle avoidance, making it more acceptable [11]. The HLPR chair (Home Lift, Position and Rehabilitation) is specialized for indoor purposes and provides functionalities such as placing the individual on a bed, providing means to access tall kitchen shelf, etc. Its main focus is to reduce back injury due to prolong use of the wheelchair. The figure below shows the chair in action. It is also used in rehabilitation centers after surgery. The chair can also support trajectory planning for independent mobility by creating a map from the sensor reading [12].

Figure 5: The Home Lift, Position and Rehabilitation (HLPR) chair [29].

Other researchers are using BCI methods to control the wheelchair. The specialized hardware and software processes the EEG signals when the user thinks of moving in a particular direction. These brainwaves are then translated and the wheelchair moves in that direction. This technology is still in its initial stages and more research is in progress. The figure below show the wheelchair Toyota is currently working on to be used for navigation using brainwaves.

Figure 6: Toyota BMI wheelchair [30].

Other researchers are combining two methods such as combining computer vision with Bio-signals to get better results. The computer vision is used for pattern detection when a particular movement is performed and at the same time Bio-signals are captured. Combining the information from two separate systems using different technique helps in increasing the confidence level of the control system and makes it more reliable [13].

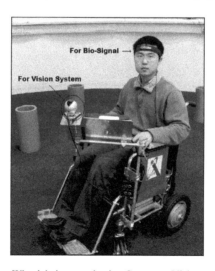

Figure 7: Intelligent Wheelchair control using Computer Vision and Bio-signals [13].

The Smart wheelchair provides three different levels of control strategies, deliberative, reactive and user-initiated. The user initiated level is used for doorway and hallway navigation. The other levels provide path planning and local obstacle avoidance using IR proximity sensors and laser range finder [14]. Implementing a particular algorithm requires a wider knowledge base of the available methods for using specific signals. A better algorithm can be written when the expected system output is known. It helps to reconfigure the algorithm to accommodate a greater degree of control. Sometimes when parameters are changed the performance deteriorates and it needs to be tested thoroughly to maintain the system stability. This is an iterative process and takes time to reach acceptable level of accuracy. Ongoing research activities are focused to increase precision and make the system more robust for real world implementation.

2   Goals and objective

The project aim is to design a hybrid Bio-signal control algorithm for an intelligent wheelchair intended to be used by quadriplegic individuals. The thought of helping people and the community is present in every individual, waiting for the right moment. I got an opportunity to channelize this energy in my project after an incident at the London tube station, where a person with disability was having difficulty boarding the train. This incident motivated me to make use of the available technological advancement in the field of robotics for the betterment of paralyzed and other individuals with severe motor conditions. The use of intelligent wheelchair will make the individual experience mobility freedom.

Bio-signals recorded from forehead are EMG, EOG and EEG signals. These signals are obtained using a headband with three electrodes. The wheelchair controller is mainly designed for indoor usage. All the Bio-signals are collected non-invasively by using single channel electrodes that are placed on the surface of the skin. The signals are filtered, amplified, and refined to be later used as control signals for the wheelchair. To accomplish this task, the project objective is divided into six stages, raw data collection and its processing, feature extraction (a challenging stage), classification of the signal blocks, decision process, wheelchair motor control and user interface development. The other stages were testing the system components, simulate the control of wheelchair using the algorithm, make necessary modification and then finally work on the actual wheelchair. Each stage was allotted separate day slot for its completion. The algorithm is the heart of this project. The performance of a system also depends on how the algorithm is designed for optimum utilization of available hardware resources. The proposed controller would be able to move the powered wheelchair in four directions i.e. forward, backward, right and left and stop. Sensor fusion technique is used for combining the advantages of the Bio-signals to make the controls more efficient [8].

The future goal is to modify the wheelchair for outdoor implementation. Provide provisions for collecting information on a daily basis. Also provide a data management system which can process these data using distributing computing to represent it in a simplified charts and tables. This database can be accessed by doctors and researcher to exchange feedback and provide appropriate medical assistance. This will not only help monitor the individual but also aid in further modification to improve the wheelchair usability and user interface.

2.1 Project description

The project is based in the field of Human Machine Interaction. The key to this project is the hardware and software working mutually together. The algorithm forms the core, which makes the captured raw data usable as a control signal for the wheelchair. The hardware mainly consists of the powered wheelchair and the Cyberlink^{TM} headband with data acquisition box. The best place to record good quality Bio-signals is the forehead of the quadriplegic individual. The Bio-signals recorded using the headbands are given to the data acquisition box for amplification and filtering. The output of the data acquisition box is given through a serial cable to an onboard processor. The signals are then processed using the data processing algorithm which uses neural network for classification. The output of the neural network is then mapped with control command for the wheelchair. A display screen provided, shows the user interface for the wheelchair.

The controller algorithm runs on an Intel^{TM} thin client processor on the wheelchair. The final output from the algorithm is given to the DSP for driving the motor of the wheelchair. The wheelchair has an embedded DSP microprocessor for controlling direction and velocities of the motors. An in-depth working of the hardware with the software algorithm will be given in the following sub sections. A cursor program is provided by Cyberlink^{TM} to allow the cursors to be moved around the screen using the brain signals combined with forehead muscle movement without using any pointing device. Using the cursor program, user can independently decide to start or stop the wheelchair controller algorithm. The display screen implemented is a touch screen, to minimize additional pointing devices. The wheelchair controller algorithm can also be initiated by a press of a button by the caregiver using the touch screen. This implementation feature help to give the individual mobility freedom with minimum human supervision.

The project is divided into two main parts hardware and software. The block diagram below shows the transformation stages of data signals into command signals. The signal moves over five stages i.e. data collection, pre-processing, feature extraction, decision-making and device control feedback. The raw data recorded from the individual using the headband with electrodes is filtered, amplified and digitized in the data collection stage. The digital signals are then sent to the on board processor where the further processing, data segmentation, feature extraction and decision making is carried out. Once the decisions generated and mapped they are send to control the wheelchair motors. A control feedback to the system is provided to increase accuracy. The display screen shows the current control command in action as a visual feedback for the individual.

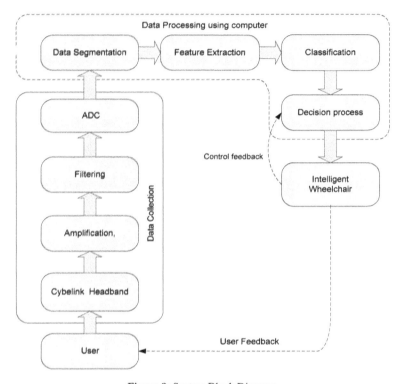

Figure 8: System Block Diagram.

2.2 Hardware

The hardware requirements of the project are the Cyberlink^{TM} headband with data acquisition box, the wheelchair with an on board thin client PC and a DSP unit connected to the motors. The following sub-points give details of the required hardware in detail.

2.2.1 Cyberlink^{TM} headband

The Cyberlink^{TM} headband as shown in the figure below has three electrodes attached. These electrodes form a single channel for data recording. The center electrode is used as a reference for the neighbouring two electrodes. Special Silver Chloride plated sensors are used which don't require conductive gel between the skin and the electrodes. Only in a few exceptional cases when the individual's skin is too dry would we use conductive gel. These electrodes require less maintenance. To get better results they can be removed and cleaned after a prolong use. The back of the headband has a Velcro strip to secure it in place. This makes it easy to adjust the headband to the individuals's needs. The electrodes on the headband record the signals which are generated due to facial muscle activity, eye movements and also brainwaves. The signals are technically called EMG, EOG and EEG respectively.

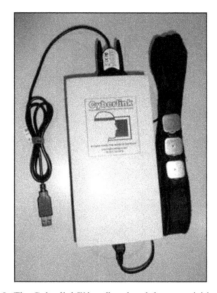

Figure 9: The Cyberlink*™* headband and data acquisition box.

The human face being made of multiple muscle groups can generate many distinct movements. These movements can be tapped by placing electrodes at key location. Sequence of different facial movements were tried and tested during the initial phase. The features which were easy to perform by a quadriplegic individual and at the same time which were high in signal energy were selected. Some of the important movements like forehead frowning has more prominent signal energy and is used as control command for the wheelchair. The forehead frowning (raising of both the eyebrows) twice is used to move the wheelchair forward, the left or right gaze is used for left or right turn commands, rhythmic jaw clenching movement is used for the reverse command for the wheelchair and the forehead frowning once is used as the stop command.

The frequency range of the EMG signal is between 70-1000Hz, EOG signal ranges between 0.2-3Hz and the EEG signal has a range of 0.5-45Hz. The signals recorded are in millivolts range and are passed to the data acquisition box for amplification [13].

2.2.2 Data acquisition box

The forehead is a convenient location for non-invasive measurement, rich in a variety of Bio-potential signals. The signals detected by the electrodes in the headband are sent to the Cyberlink^{TM} data acquisition box. The box is powered using two AA batteries. The box can be easily connected to a computer's serial port. It contains a Bio-amplifier and signal processor inside for making the signal understandable for higher level processing. The forehead signals are amplified, digitized and translated by a decoding algorithm into signals of multiple frequency bands. There are twelve different signals bands created using the recorded EMG, EOG and EEG signals. The final outputs are one dimensional discrete signal. The conditioned signals are then transmitted to the computer. The data acquisition system has the following features [8].

- Analog Gain: 200,000.
- Analog Bandwidth: 0.2 Hz to 3,000 Hz.
- Analog to Digital Conversion: 6 Channels, 12 Bit accuracy.
- Isolation: 2,500 Volts.

The Bio-signals strength is highly variable as it depends on number of factors such as size of the electrodes, contact between the electrode, the skin resistance, the thickness of the tissue through which the signals pass, the time for which muscle action potential is high, etc.

The EMG signal available from the electrodes is around 5mV. As the human body acts as an antenna, it picks up strong signals about 2V from the surrounding environment. As a result it becomes difficult to distinguish between the actual signal and noise. To prevent this we make use of reference electrode. The reference electrode is the center electrode of the headband. As the noise is same through the body we subtract the reference electrode signal from the other two electrodes and get background-noise reduced signal. This is done using differential amplifiers and the method is called Common Mode Rejection Ratio (CMRR). The data acquisition box passes the signal through a low pass filter to remove irregular peaks of high frequency component. The box uses filters to pass required frequency band and reject unwanted frequencies. For EMG and EEG signals Butterworth filter is used as it has smooth pass band and stop band. For EOG signals we use Elliptic filters due to its sharp cut-off features. The final output signal is digitized using ADC [15] [16].

When lights hit the retina of the eye an electrical dipole is formed within the eye. When an individual glances either left or right the dipole shifts towards one of the electrodes and leads to a potential difference. The recorded potential is the EOG signal. The signal amplitude is around 10-100mV. EOG signal amplitude mainly depends on the position of the eyeballs with respect to the conductivity of the environment. For better usability of the signal there should be proper amplification and elimination of interference due to noise [17].

The EEG signal is the recording of the electrical activity of the brain. In this project the signal is recorded from the forehead region. Normally EEG is recorded from the scalp region using electrode cap. The signal is created during synaptic activity when millions of neurons are triggered together. The human brain has billions of neurons which forms a complex neural network structure. The figure below shows the key elements in a synaptic transmission. The Axon carries the spike of 'Action Potential' to the next neuron. The 'Action Potential' is caused due to difference in potential around the neural membrane. Strength of the signal, weakens as it passes through a number of layers before it is recorded by the electrodes. The signals recorded by the headband are limited, as the signals are recorded from the forehead region using a single channel. With the advancement of technology in the past decade EEG signals are used for BCI applications. The Bio-signals recorded each time are never entirely identical. The signal strength changes between individuals, time of the day they are recorded, body movements and due to external interference.

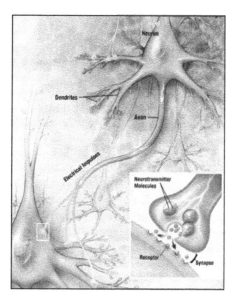

Figure 10: Action potential travels along the axon of a neuron [31].

2.2.3 Wheelchair

Nowadays wheelchairs are made intelligent by coupling them with embedded computing power and sensors. Such chairs are called intelligent wheelchairs. The intelligent wheelchair shown below is the 'Robochair' jointly developed by the University of Essex and Chinese Academy of Science in 2004. This chair is used in the project, with selected features. From the features below the webcam is not used for this project. It has the following specification

- Six ultrasonic sensors for obstacle avoidance.
- DSP TMS320LF2407 controller chip for the two differentially-driven wheels.
- A joystick controller connected with the DSP for control override.
- A Logitech 4000 Pro Webcam for recognizing the user's head gesture.
- Intel Pentium-M 1.6G Centrino with Windows XP operating system and works on 24V batteries.

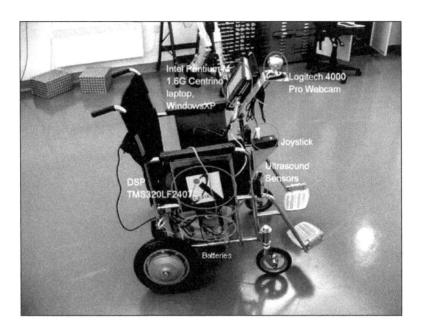

Figure 11: Picture of the smart wheelchair (RoboChair) for the project [18].

The RoboChair has two control modes. Irrespective of the mode, the data collected by the sensors from the surrounding environment are sent directly to the DSP for obstacle avoidance.

Intelligent mode: In this mode the wheelchair obtains control commands from the computer. The computer processes the data received from the sensors and processes it to make a decision based on the available input data from the user and the other additional sensors. Finally, the control decision is send to the DSP controller that actuates the final motor actions.

Manual control mode: In this control mode, RoboChair is controlled directly by the joystick connected to the DSP controller. This mode is already built in the RoboChair. This mode is better for corridor navigation or if the motors turn inappropriately towards a particular direction and need a manual intervention. The obstacle avoidance is done independently using sonar sensors. In manual mode the user can perform reserved states that are not allowed in the intelligent mode.

The TMS320LF2407 DSP chip offers great processing speed as it uses pipelining and multithreading concept. Compact peripheral integration providing 30 MIPS and the programs take up less space. The results in high driving performance due to better real-time signal processing. The obstacle avoidance module is embedded for safe operation of the controller at crucial periods [18].

2.3 Software

The amplified and filtered data from the data acquisition box is displayed on the computer using Cyberlink^{TM} Brainfinger software. The software shows eleven frequencies bands for the EOG, EEG and EM|G signals recorded. The signals that are splitter into eleven frequency bands are displayed using the software as shown in the figure.

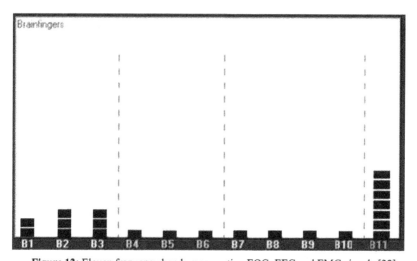

Figure 12: Eleven frequency bands representing EOG, EEG and EMG signals [32].

The three bands from B1-B3 represent right-left glance. The bands goes up when the user is either glancing left or right and goes down when looking straight. The bands from B4-B6 indicate the Alpha frequency which goes up when the person is in calm state. Bands B7-B10 indicates beta frequencies of EEG signal which rises when the user is concentrating on something. The B11 band represents the EMG signal activity. This band rises when facial muscles are used for triggering a command.

The software has a panel to display each of the signals separately as shown in the figure below. The signals are in the following sequence EEG, EOG and EMG.

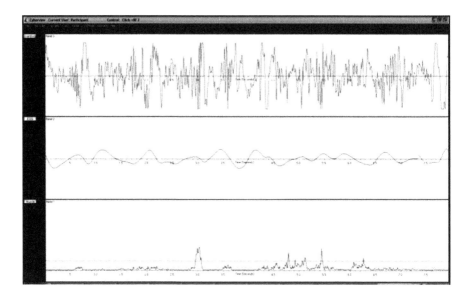

Figure 13: EEG, EOG and EMG signal displayed using Brainfinger software.

2.3.1 Software description

Along with the data acquisition box and headband, Brainfinger software code is provided by Cyberlink^{TM}. This code is used as the basic platform for implementation of the project. The code is reused to obtain the signals captured by the headband. Further addition in the code is written using Visual Studio C++ platform. A separate program was written for collection of data for the training of the ANN. The program captured the values of EMG, EOG and EEG signals and saved in a file. Another program was written to process the saved data. This program is used for feature extraction and providing the values to the ANN. The output of the ANN are mapped with control command and are used as the input to the decision logic. This block then sends necessary commands to the motors for respective movements of the wheelchair. The flow chart briefly shows the steps taken in the software algorithm to control the wheelchair.

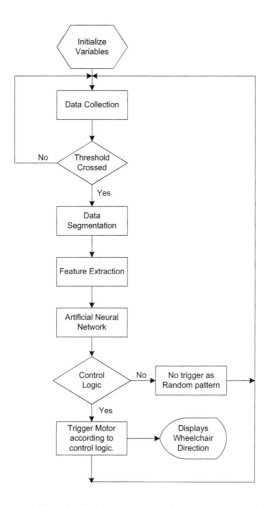

Figure 14: Flow chart of software stages to control the wheelchair.

2.3.2 Data collection

In this stage the data signals collected from the data acquisition box are saved in a file for further processing. The sampling frequency of the device is 100 Hz, as 100 samples are obtained in one second. The data is saved in a file, column wise with three rows containing EMG, EOG and EEG data. Initially we need to collect specific set of sample data for feature extraction process. There are four movements that will be used for controlling the wheelchair.

The control movements are:

Forehead Single Click (FSC),

Forehead Double Click (FDC),

Rhythmic Jaw Movement (RJM),

Left Eye Glance (LEG),

Right Eye Glance (REG).

The above movements are selected as they are unique, easy to extract from other signals and easy to learn. EMG signal are generated when we do FSC or FDC. One time lifting of both eyebrows is called the FSC. Lifting the eyebrows twice in a short interval is the FDC movement. EOG signals are generated by LEG and REG movements. They involve glancing at left or right side for a small moment [13]. EEG signals are used in this project to add their features in the wheelchair control method. These signals are obtained and captured when we are doing actual action or thinking of moving any of our limbs. The EEG signals are recorded at the same instance we record EMG and EOG signals. The advantage being that the EEG signals are then correlated corresponding to the specific movements as they are capture. This correlation between the three signals is used to reduce the effects of noisy face movements. As the noisy movements can disturb the normal working of the system when the user is talking or looking around.

The Cyberlink^{TM} device has a sampling rate of 100 samples per second. Each sample contains data of all the three Bio-signals. Each training sample of the training file is made of 35 seconds consisting of seven movements. Each movement is done once in a five second time frame. In the earlier stage of training it was decided to use Rhythmic Jaw Movement (RJM) (jaw clenching movements) for 'move backwards' or reverse command for the wheelchair. But during actual implementation the clashes between the FSC, FDC and RJM movements were noticed and the RJM movement was not used as a control command. During the training the movements used in sequence are relax, FSC, REG, RJM, LEG, FDC and again relax state. In the relax state the individual can look around, whisper, smile and do movements other than the five reserved movements.

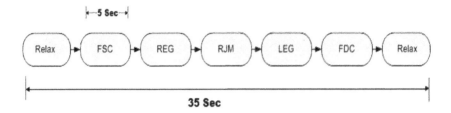

Figure 15: Different states in a training sample data.

When the user is not wearing the headband, the impedance between the electrodes is high. Once the headband is worn by the individual, the signals needs some time to return to normal levels. The relax state is present at the start for this reason. The relax state in the end is used to capture signals which are random. The neural network can be trained to ignore these random signals during its training stage. The states of the control command are so planned that no two signals of the same Bio-signal type are adjacent. This arrangement will make the data easy to separate in the next stages even if there is some degree of overlap. The data in the above format is saved in a text file. A total of 150 data files are created, which consists of 130 training sample and 20 test sample data.

2.3.3 Pre-processing and Data Segmentation

The data collected and saved in the previous stage cannot be used directly for data segmentation. There were some points which were not considered while saving the file. When the file was saved each data signal was followed by a comma and then by space. The comma and space have to be removed from the data to be used in data segmentation stage. After getting the signal data in required format, then they are split into 500 samples, as each state of the training sample is five seconds long it equals to 500 samples per state. After this stage the data is ready for segmentation. Data segmentation is done using threshold based method in time domain. Two threshold operators are used to select the data in range from the stream of sample data signals. The thresholds values are decided by observing the signal range during which it varies by a greater proportion. In data segmentation stage the range of data selected is kept wider compared to the next stage of feature extraction. The general equation for threshold detection is written in (1). Where TH is the higher Threshold value and TL is the lower threshold value. The output value of the function is 1 if the value falls in the range else output is zero. The EMG signal varies from 0 to 2047, the EOG and EEG signal vary between -512 to 512. The EOG signals from right and left eye is distinguished using the phase of the signal. These signals are then saved for further processing.

$$y[n] = \begin{cases} 1 & when \ TL < n < TH \\ 0 & else \end{cases} \qquad (1)$$

2.3.4 Feature Extraction

Feature extraction stage is the challenging stage. The features extracted from the signal should be sufficient to represent the movement which triggered it. The data obtained from data segmentation stage is used to narrow the search to select the features at the precise moment the signals are triggered. The amplitude or phase thresholds of the signals are further reduced to extract features from the data samples. Once the data range is available three features are extracted from the samples. The features extracted are Absolute Mean Value (AMV), Root Mean Square value (RMS) and Average Crossing value (AC). The extracted features are then saved in a file for training the artificial neural network.

In the project both quantitative and qualitative methods were implemented. During data collection the theoretical knowledge of Bio-signals was applied for selecting the signals. The qualitative approach was used to find the thresholds for each pattern, for appropriate control of the wheelchair. For FSC recognition a gradient function $\nabla f(n)$ from the EMG signal $f(n)$ is calculated which gives the deviation value at the nth sampling. N is the total number of samples. The equation is as shown in (2).

$$\nabla f(n) = \begin{cases} f(n) - f(n-1) & n = 2, ..., N \\ 0 & n = 1 \end{cases} \qquad (2)$$

The $f(n)$ value is directly propotional to the tension in the muscle. The value of $f(n)$ varies between 0 to 2047. If the value obtained is more than the range due to heavy face movement it is reduced to 2047. Single click is detected by calling a function which checks the status of EMG signal. If the status flag is set, single click is identified. As FDC is achieved by two consecutive FSC. A new trigger is used for FDC, which is set when two FSC occur within a short interval of time. The FSC and FDC are used as 'Stop' and 'Move forward' command for the wheelchair respectively [13].

34

For left and right eye glance recognition state funtion is defined in (3)

$$K(n) = \begin{cases} 1 & if \{ n: g(n) \geq h1, n \in \Delta T \} \\ -1 & if \{ n: g(n) \geq h2, n \in \Delta T \} \\ 0 & otherwise \end{cases} \quad (3)$$

Where, $K(n)$ is the state function, $g(n)$ is the EOG signal and $h1$ and $h2$ are user dependent thresholds for EOG signals. When $K(n) = 1$ REG is identified and when $K(n) = -1$ LEG is detected, else the state function is zero [13].

Feature fusion technique is applied to the Bio-signals to provide reliable control for the powered wheelchair which helps in discarding noisy facial movements. Features extracted from all the movements are used together to make a final decision as shown in the figure.

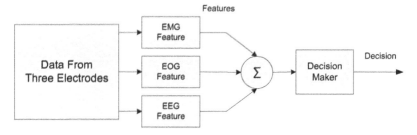

Figure 16: Feature fusion for better movement discrimination.

2.3.5 Classification

he human brain can identify a pattern from a set of data quite easily. Consider one of the problems faced by engineers while designing systems which require identifying a person from a picture or a video stream. We as humans can see the all the features of the person and will identify the person in couple of seconds. If the same task is given to a computer, it will first bread down the picture in small parts. It will then compare these small parts of the picture one by one with a reference picture and after complete patter matching it will give final result. If high accuracy is needed it will reduce the size of the region, it compares with a reference picture and then repeat the same steps. This consumes a lot of time and processing power. If the total time required is to be decreased the system accuracy reduces and leads to faulty results. This problem is encountered in identifying two similar pictures taken at the same time. The complexity just escalations multi fold when a person is to be accurately identified among different pictures or a live video stream.

My project involves classification of complex Bio-signals in real-time. The response time of the entire system has to be fast and also accurate every time, as the safety of the individual depends on the system algorithm. A computer cannot work with the same efficiency as there are no perfect algorithms to map the features and to correlate the data the way humans can. The solution for this problem is to simulate the human brain using computational algorithms. The artificial neural networks (ANN) are adaptive computational algorithm based on biological neural network. They compute data in parallel, using small processors which are interconnected together.

In the research world, ANN is implemented to solve complex challenges in distributed data computation, pattern classification, function approximation, forecasting and also to solve problems which are difficult for conventional computers or humans. ANN cannot be programmed to solve a particular problem, it learns a function by learning the examples of that function as to how the inputs and outputs are related with each other. All ANN have a specific application, learning rule, architecture and a learning algorithm. The learning rule can be supervised, unsupervised or hybrid. The learning algorithm designed is dependent for a particular ANN architecture.

Classifiers are used to distinguish and select different patterns from available input signals. In this project artificial neural network is used as the classifier. They are made of simple elements called neurons and were inspired by understanding how the human brain functions. As shown in the figure below a basic network has three layers i.e. input layer, hidden layer and output layer. There can be more than one hidden layers between the input and output layer.

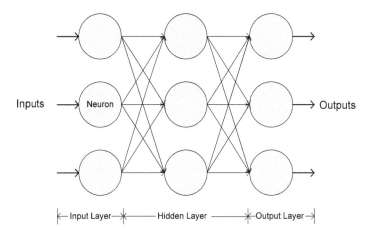

Figure 17: ANN with three input neurons, a hidden layer and four output neurons [20].

The input and the output layer take in the values of the function, we need to approximate. A function is defined by the number of neurons and the connection values or weights between them. The connection and the number of hidden layer give an idea about the function's performance. More hidden layers and neurons are required to solve complex problems but too many or too few neurons may degrade the overall performance.

On a basic functional block level of ANN, the output of each neuron is a function of the weighted sum of the input and a bias value. The bias is much like weight, except it has a constant value. During the training process, the bias value is change accordingly after every training sample to get an optimum output. The learning process is iterative in nature and also dependent on the training samples provided to the algorithm.

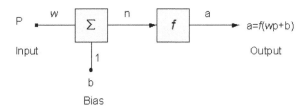

Figure 18: A weighted neuron structure with bias [19].

The output of an ANN is the computational value of all the neuron outputs present in the network architecture. To determine the strength of the neuron's output, the result is given to an activation function. The most common activation function is the sigmoid function, which provides outputs between 0 and 1.When the neuron is in training stage, it slowly adjusts the weights of neurons in the network to approximate the output for a given input. After training the network with sufficient training samples , it should give approximate outputs even for samples it was not trained [19], [20]. The neural network identifies a close match to the patterns presented to it and then only explores others patterns that are similar. It tracks down a number of possibilities and finally selects the most promising match. The result is then propagated to the next stage where the command is given to the chair. The same procedure is carried out repeatedly for different iteration value as new inputs are presented to the network.

In this project the ANN has nine inputs neurons, one output neuron and fifteen neurons in the hidden layer. One neuron is assigned per input. The inputs to the ANN are the features extracted from the data signals. The neural network architecture is based on Back-propagation algorithm. There are 9 neurons in the input layer. The hidden layer has three sub-layers each with 5 neurons. The output layer has 1 neuron. The ANN is simulated using FANN Explorer software which is included with FANN library.

2.3.6 Algorithm

Incremental training algorithm is applied for this system and it is a standard for Back-propagation algorithm. This algorithm changes the weight of the neurons by propagating the error backward from the output layer to the input layer while adjusting the weights. In this method the weights are updates after each training pattern. Error function selected is linear and the learning rate is set to 0.1. The learning rate determines the assertive level of the training. Learning momentum is set to 0.4. It is used to boost the training algorithm. The normal range is 0 to 1. If the training momentum is set too high, it can degrade the training process. The activation function used for the layers is sigmoid symmetric as its output ranges between -1 to 1. The output range is used to distinguish the four patterns at the output layer and also make training more precise and faster. The output layer has one neuron which has four states i.e. -1, -0.5, 0.5 and 1. These signals are given to the control logic to trigger motors of the wheelchair. The activation steepness is set to 0.5. It's the step change in the output. It is indicates how fast the function changes from minimum to maximum state [21].

The bias and weight in the neural network are changed using the Rprop (Resilient backpropagation) algorithm. The basic principle of Rprop is to remove the harmful effects of the size of the partial derivative on the weight step. More detail can be found in (Riedmiller and Braun, 1993). The following pseudo code section indicates the RROP adaptation and learning process. The minimum (or maximum) operator is supposed to return minimum (or maximum) of the two numbers. The sign operator returns +1 if the condition is true, -1 if the condition is false and 0 otherwise [22].

$$\forall i, j: \Delta ij(t) = \Delta 0$$

$$\forall i, j: \frac{\partial E}{\partial wi, j}(t-1) = 0$$

Repeat

$$Compute\ Gradient\ \frac{\partial E}{\partial w}(t)$$

For all weights and biases {

$$if \left(\frac{\partial E}{\partial wi, j}(t-1) * \frac{\partial E}{\partial wi, j}(t) > 0\right) then\{$$

$$\Delta i, j(t) = minimum\ (\Delta i, j(t-1) * \acute{\eta}, \Delta max)$$

$$\Delta wi, j(t) = -sign\left(\frac{\partial E}{\partial wi, j}(t)\right) * \Delta i, j(t)$$

$$wi, j(t+1) = wi, j(t) + \Delta wi, j(t)$$

$$\frac{\partial E}{\partial wi, j}(t-1) = \frac{\partial E}{\partial wi, j}(t)$$

}

$$else\ if \left(\frac{\partial E}{\partial wi, j}(t-1) * \frac{\partial E}{\partial wi, j}(t) < 0\right) then\{$$

$$\Delta i, j(t) = maximum(\Delta i, j(t-1) * \acute{\eta}, \Delta min)$$

$$\frac{\partial E}{\partial wi,j}(t-1) = 0$$

}

$$else\ if\left(\frac{\partial E}{\partial wi,j}(t-1) * \frac{\partial E}{\partial wi,j}(t) = 0\right) then\{$$

$$\Delta wi,j(t) = -sign\left(\frac{\partial E}{\partial wi,j}(t)\right) * \Delta i,j(t)$$

$$wi,j(t+1) = wi,j(t) + \Delta wi,j(t)$$

$$\frac{\partial E}{\partial wi,j}(t-1) = \frac{\partial E}{\partial wi,j}(t)$$

}

}

Until converged [22]

2.3.7 Training and Testing

Once the network weights and biases are initialized, the network is ready for training. The network is trained using 130 training samples. One cycle after which the weights are adjusted to match the output in the training file is called an epoch. The weights of the neural network are changed after an epoch. In the project, maximum number of training epochs is set to 1000 and the desired mean error is set at 0.001. While training the neural network tries to achieve minimum error by calculating the least mean square error value. The Riedmiller and Braun algorithm as explained in [22] is used for faster convergence and simple implementation. Training stops when either the numbers of epochs are reached or if the desired error value is reached. A single training file is made for the ANN containing 320 feature extracted samples each with nine data value. After training if the error is high the weight, bias and structure of hidden layers are changes to get better response. The figure below shows the mean error during intermediate iterations. A total of 43 iterations were performed to get a final acceptable mean square error plot. The ANN is later saved to be used for test samples. The plots are generated using the GUI of FANN library and digital pictures of the plot are taken for demonstration.

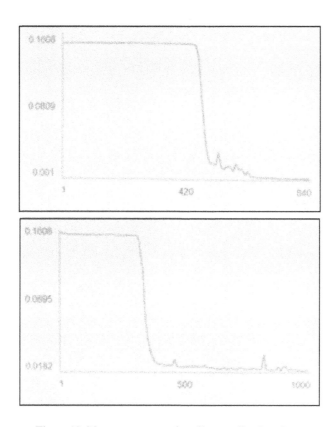

Figure 19: Mean square error plot of intermediate iterations.

After the training is complete, 20 more samples are collected for testing the ANN. First sample data for 10 test control command is given to the ANN to see if the training was successful. After getting the outputs again 10 test samples are given to the ANN. The outputs obtained from the ANN are then mapped in the next stage. In the later stages real time data is given to the ANN and the output is observed for the algorithm.

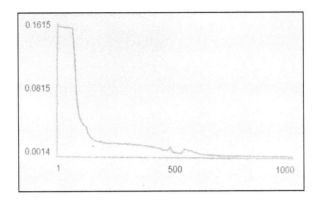

Figure 20: Mean square error of the finalized ANN after training.

2.3.8 Pattern Mapping

The output of the neural network is of float data type and has to be standardize to respective patterns numbers. A graph is obtained by plotting the output of ANN for the 20 test data samples. The plot helps to find the range of output that needs be regulated. On the x-axis we have the control patterns which are given to decision logic to trigger the wheelchair motors. On the y-axis, output of the neural network is plotted.

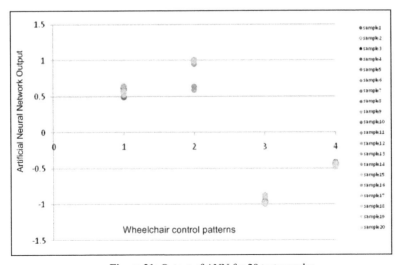

Figure 21: Output of ANN for 20 test samples.

From the graph the maximum and minimum possible values are obtained for each patter to a relevant ANN output for mapping. The values 1, 0.5, -0.5 and -1 indicate the four pattern FSC, FDC, LEG and REG respectively. As we can observe from the graph the ANN has some troubles distinguishing between the FSC and FDC pattern. One of the reasons is that in the algorithm more priority is given to FSC then FDC, for safety of the individual as it represents the stop command.

Furthermore FDC movement is repetition of FSC separated by a small time delay. This time delay has to be the same each time the command is performed. If the time delay is more the FDC command will not be accepted by the system and the wheelchair will stay standstill. The mismatch between the two commands does not hamper the control. The control algorithm was designed for this response as stopping the wheelchair is more crucial. In this stage of the project the FDC command has to be repeated a few times before it get detected. This error can be reduced by training the ANN with more training samples.

2.3.9 Decision process

The mapped outputs of the ANN are used to make the final decision to trigger the motors of the wheelchair. The control decision logics are as shown in the table below.

Trigger Movement	Control Command
FSC	STOP
FDC	MOVE FORWARD
REG	TURN RIGHT
LEG	TURN LEFT

Table 1: Wheelchair control decision logic.

The decision logic is used to map the trigger movements with control commands for the RoboChair. The sequence also indicates the importance given to each trigger movements. Stops command is given more preference than any other movement for safety reasons.

In the algorithm only after FDC command is performed, LEG and REG command will be executed. If any other pattern is detected other than FDC, the wheelchair is not given any control command and it remains standstill. This clever feature was implemented to avoid the wheelchair to turn when the individual is just casually glancing left or right without actually commanding the wheelchair to turn. As a result LEG or REG commands can only be executed after FDC command, confirming the individual real intentions of turning the wheelchair.

2.3.10 User interface

The touch screen monitor attached to the wheelchair is used by the user to interact with the system. When the program is loaded the pop up screen appears as shown in figure below.

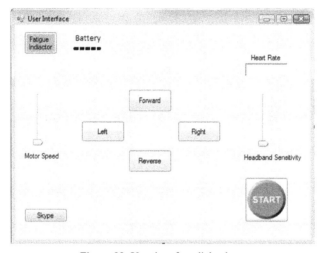

Figure 22: User interface dialog box.

The user interface design also includes features for future work. During the initial stages it was decided to add these features in the user interface but due to time constraints they were not implemented in the system. The final user interface had four buttons to indicate the four directions, one start button and one exit button to stop the program. Once the start button is pressed the program starts reacting to user inputs. A specific directional button gets highlighted when the wheelchair is moving in that particular direction giving a visual feedback. Once the exit button is pressed the program stops capturing all the bio signals. The program is started by pressing the start button at the lower right. The start button has a toggle behavior similar to a virtual music player interface. When the audio stream is activated by pressing the Play button, the Play button changes to Stop button and vice versa. The four display buttons in the center are used to indicate the current movement status of the wheelchair. The other features are for future implementation. The track bar on the right is used to adjust the sensitivity of the headband electrodes. The one on the left is used to change the speed of the motor as required by the user.

There is a fatigue indicator at the top left which changes color from green to red when the signal peak goes below a threshold value for a specific number of times or the wheelchair is used continuously for more than 10 hours. This indicator is used to prevent false triggering of commands when the user is exhausted. Next to the fatigue indicator there is a battery level indicator, which displays the power of the onboard batteries which powers the entire wheelchair control system.

Real-time heart rate is displayed which indicates the need of oxygen by the body and also help in tracking other medical conditions. The Cyberlink^{TM} device has an add-on accessory for monitoring the heart rate. A application interface button is provided for communication. In this case Skype^{TM} (a VoIP service) is used to provide communication services. With the availability of high speed Wi-Fi, the on board PC can be easily connected to the internet. A communication tool is provided to the user to interact with other people and not feel isolated. The start button can be further programmed to save all the biological data obtained during the usage of the wheelchair. The data collected can then be uploaded to a medical database for doctors and researchers to analyze and help in further medical assistance for the individual.

3 Evaluation

The evaluation of the system is done in three steps, testing individual components of the system, sub-system testing and testing the entire system as a whole.

3.1 Component testing

The two main components which require testing are the Cyberlink^{TM} device and the RoboChair wheelchair. The device works on two AA batteries for around 25 hours. The general troubleshooting for inactive system response, is replacing the batteries. The system can also be plugged in directly to the on-board batteries.

In the headband, electrodes are firmly attached to reduce noise due to electrode movement. After prolonged usage, electrode needs to be cleaned to improve signal strength.

There are few prerequisites that are required before the system can be operated. When the headband is not worn on the forehead there is high impedance between the electrodes, causing abnormalities in the signals amplitudes. When the individual wears the head band the signal takes some time to reach its normal level. As a result, after wearing the headband the individual needs to wait for few seconds before initiating any control commands.

The data acquisition box has an input from the headband and an output to the computer through a serial cable. The headband and data acquisition system are checked by using them with the Cyberlink^{TM} Brainfinger software. The software displays all the captured signals on a signal panel interface. The signal are active depending on their types, EEG fluctuates constantly, EMG varies with facial muscle movements and EOG signals varies as the individual glance in different directions. An error message is displayed if the device is not connected properly.

The various Bio-signals are displayed on the panel only when the device is connected properly and is recording the signal correctly. The software also checks the availability of the device on all the ports and then gives the error message if the headband is not detected. This method is applied for situations when a Serial-to-USB convertor is used and is plugged into different ports. The wheelchair is tested by switching on the on-board computer and the wheelchair power switch. The wheelchair is powered by two rechargeable 12V batteries. The batteries need to be charged after every use. The RoboChair is switched into manual mode and is operated using the joystick. The headband is worn and checked for connectivity. The Bio-signals are observed on the displayed interface. Once all the hardware components are tested, system software modules can be tested.

3.2 Subsystem testing

Subsystem testing involves testing system software algorithm in parts. The algorithm for data collection is tested by executing the data captured part of the code and the results are saved in a file. The file is later checked for changes in the sampled data for specific time interval where the control commands were initiated.

The next part is to test the data processing algorithm. The entire file generated in the previous step is read by the processing part of the code as an input to its functions. The algorithm has different sub functions that are checked by using break points in the IDE. The output obtained from the code is checked with the excepted output. Necessary changes are made if the code has exceptions or errors during execution. All the pre-conditions for accepting the control command are verified. ANN output is checked for respective control command output. It has to only pass the selected control command to next step, all the other random noise movement signals have to be rejected. Once the processing code is working perfectly the decision logic code is tested for its specific output that are given to the wheelchair motor for the final direction command.

3.3 System testing

All the program code is integrated together and the entire system is tested. Testing of the entire system is carried out to find any defects from any unanticipated interaction between the subsystems. If there are any bugs or errors, they are located and rectified. The system is tested again to find any other bugs. This process is iterated until acceptable results are obtained [8]. In this testing the processing code gets the input directly from the data collection program in real-time. The data is collected for every two second and processed to find trigger movements. The system output is tested using different methods, the output of the system is observed for valid and invalid inputs. The algorithm is also tested during different times of the day and night. It was observed that the signal strength is high in the morning and thus the detection of the control command.

The algorithm is also tested using a test map. The system performance is obtained by taking some test runs and observing the results. For testing the output, the control commands are simulated using arrows for direction of the wheelchair.

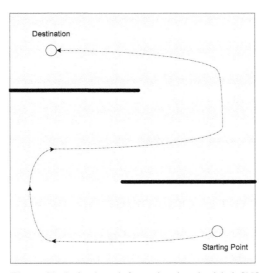

Figure 23: A simple path for testing the wheelchair [32].

The system had a slight problem distinguishing LEG, REG and random signals. The error were reduced by training the ANN for more data samples. A simple map for test simulation is made as shown above to imitate a path.

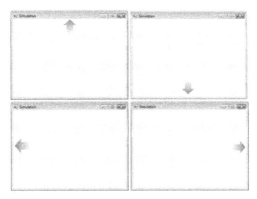

Figure 24: Simulation of wheelchair movement using directional arrows.

Initially the test is carried out by simulating directional arrows on the screen. The arrows displayed on the screen indicate the direction of the wheelchair movement. The black lines represent obstacles in the path of the individual through which the wheelchair needs to be maneuvered. The system is initially tested for 10 runs with all the control movements. Later more training to the ANN and slight modification in the code the system is tested for 10 more test samples. During evaluation it is checked if the pattern is detected correctly in response to the control command. Final outputs, displayed as arrows on screen are observed to find the error in the system. The test result of the initial 10 test run is as shown.

For 10 runs	FSC	FDC	REG	LEG	Random (noisy movement)
Correctly detected	10	9	8	3	4
Wrong detected	0	1	2	7	6
Correct final output	9	8	5	2	2
Average Time taken (Sec.)	2.3	2.7	3.5	3.7	3
Performance (%)	90	88	63	66	50

Table 2: Test01 Result.

The performance and the average time for each movement are evaluated. The time delay between the command and the actual movement of the wheelchair is little more than two seconds.

For 20 runs	FSC	FDC	REG	LEG	Random (noisy movement)
Correctly detected	20	17	19	18	12
Wrong detected	0	3	1	2	8
Correct final output	19	14	13	12	6
Average Time taken (Sec.)	0.7	1.8	2.7	3	3
Performance (%)	95	82	68	67	50

Table 3: Test02 Result.

During testing movements like heavy jerks are avoided. When random facial movements are given to the wheelchair, the commands are not accepted and it remains standstill. This feature is tested for its functionality. There are only few facial movements like the RJM that can interpreted as a false trigger command. To improve the system performance more training samples for LEG and REG are collected and the neural network is trained for more training data. The trigger level for FSC and FDC were marginally reduced which improved the final output detection. The results of the second test are as shown below with 20 runs. The performance of the system increased but the progress was not that significant. Few more iterations are required to reach a better level of accuracy.

After complete evaluation of the system, the ANN is trained for more samples and the algorithm is modified to be more responsive. The sampling frequency is increased and the motor control code is added to the Algorithm. Once the motor controls are working the entire system is tested with the wheelchair. The system is tested with real-time control command from the headband and the motors are observed for the direction corresponding to the control command.

4   Future development

In relation to Moore's law with computer processors double in complexity every two years and with new technology developing rapidly. In every project there are always some features that can be added to the system to make it more versatile and future proof. The features that can be added to the system are mentioned in the following subsection.

4.1 ANN training:

The ANN was trained with 130 training samples during the training stage of the project. The mean error value obtained was acceptable but during testing stage there were few sample data patterns which were not identified correctly. As a result during actual implementation there were some errors in distinguishing between patterns. This can be reduced by training the network for more samples especially 80 more samples with a lower tolerance for noisy facial movement, to discard them when detected.

In the pattern matching step we have seen ANN has some troubles distinguishing between the FSC and FDC pattern. As both the movements are separated by a small time delay, this should be studied and set according to the individuals operating the wheelchair. Individuals who are physically weak the time delay will be greater compared to a healthy adult. The time delay can also change due to fatigue. All this parameters should be taken into account in the user interface. Additional feature can be added to the user interface, a side bar to adjust the time delay manually or automatic, it can be calibrated according to the individual's requirement.

During the training of ANN all the control commands were given together. I have an impression that if we only train the ANN for once specific control command. This can make the ANN differentiate between commands in a better way. The major concern will be separating all the training files for specific command and then training the ANN. Furthermore for increasing reliability collecting the entire possible signal pattern from a user would be a time consuming process but rewarding at the end.

4.2 Extra features in GUI:

Features such as fatigue indicator can be implemented. The signal amplitude can be used to detect if the user is exhausted. The signal expands in the time domain and shrinks in the frequency domain due to the effect of fatigue has been found by researchers. The fatigue indicator checks the peak of the incoming signal with the amplitude of the previous signal. If for the same command the signal falls below a threshold value repeatedly then the fatigue indicator is activated. The indicator is turned on even if the user operates the device continuously for more than 10 hrs. This helps to avoid misinterpreting a command.

A heart rate monitor can also be added to monitor the level of physical exertion and respiration rate. Similar wireless-enabled embedded modules can be used to obtain multiple Bio-data signals in real-time. The biological data obtained can be collectively saved on a database for the doctors and caregiver to monitor the individual. Data management software can be programmed to continuously map all data and only sent important events to the database. This will reduce the load on the data exchange servers. The data sent should be represented in more meaningful way using data visualizing techniques. This will help doctors be more productive rather than spend time to understand the vast amount of data.

This Continuous signal monitoring is helpful especially for individuals with degenerative muscle condition. As the condition gets worse, a better sitting arrangement can be provided for the individual. Furthermore an intelligent close loop system is possible which self-adjust different parameters for the individual, from the meaningful data recorded. Minimizing the waiting period required for analyzing the results.

A communication service can be added with the system for the user to be in touch with family members and friends. With Wi-Fi availability Skype^{TM} as a VoIP service can be added to the system. It being a free service can be added to the system with no additional cost. The individual can add a communication service of their choice which best suits their needs. The service providing software should be low on utilizing system resources available on the wheelchair. It should be set on a lower priority as it can consume more processing power than the actual control algorithm.

4.3 Headband:

On the hardware side the number of channels used should be increased. The electrodes used for detecting EOG signals should be placed away from the central region of the forehead for better signal recognition when the user glances left or right. Instead of a single channel, multiple channels can also be implemented for the system.

As a future implementation we can completely eliminate the cable between the headband and the data acquisition box. The signals can be transferred wirelessly over a secured channel using WBAN (wireless body area network) technology.

4.4 Additional improvements:

Motor speed control: To improve the user comfort a variable motor speed control can be set to increase the speed of the motor progressively. As the speed of the motors is kept constant there is a small jolt in the motor when the navigation commands are triggered.

Wheel design: The wheels used for the smart wheelchair are similar to that of a normal wheelchair. Mecanum wheels can be implemented to improve maneuverability of the wheelchair in tight spaces. As it has omnidirectional wheels that move in every direction and provides smooth in the place directional change. The transitions are smooth as it is design to exert minimum torque and surface friction.

Field research: On field investigation is necessary to determine which movements a quadriplegic individuals can do easily as there are always exceptional cases which involve the use of a different approach. A user specific algorithm can be built from actual feedback from the individuals and implement them as control commands. Furthermore additional features discussed in the previous subpart can be added to the user interfaces which are more helpful to cater the various requirements of the individual.

4.5 Challenges encountered

There were plenty of software challenges during the initial stages which are discussed in this section.

Compatibility:

The program algorithm was decided to be written on VC++ platform. One of the foremost problems was that the libraries for Artificial Neural Network (ANN) available were mostly in C#. A C# library was not compatible with the C++ platform. Other libraries in C++ were searched but were not easy to implement without major changes in the code. Nevertheless the search helped in comprehensive understanding of different types of neural networks and their specific application. The compatibility issue was solved by using the GUI of FANN library for building, training and testing the sample data. The GUI was easy to use and parameters of the network were easy to modify. Graphs were obtained from the GUI FANN Explorer provided with the FANN library.

A function was written in C++ to call the network for testing and it returned the final value. Initially there were some liking errors for the library when used directly. As the library was old, it had to be compiled using Visual Studio 6.0 on a Windows XP machine, as the necessary file were not build and linked in Visual studio 2008 using Windows Vista OS. Once the file was compiled, few more lines were added to make the file compatible with Visual Studio 2008 version. It is recommended to use FANN library for fast training of the neural network.

Real-time Integration:

The program algorithm for data collection and data processing was written in different projects files for proper testing and debugging of the individual parts of the algorithm. Individually the code worked perfectly. Once these two part of the algorithm were integrating many problems emerged, specially due to real-time data processing. During training, the function reads data after every 5 second and then process it. For real-time it was decided to read the data for every second and process it. The thresholds were also changed to accommodate new values during real-time implementation, this was quite time consuming. This approach cause incorrect results due to the vast data stream available for the function. Similar problems were faced when the data was to be processed after every 2 seconds. Finally acceptable results were obtained at 3 seconds but the output of ANN was inaccurate. The thresholds were further changed, sampling time modified and the problem was solved. It is recommended to train the neural network with data samples collected for less time interval for better working during real-time data processing. The process of feature fusion and ANN algorithm also needs to be iterated till required accuracy is reached.

To improve the accuracy of the system the signals were observed again using the Brainfinger software for each control movements. It was observed that when a signal is triggered it affects the signal amplitude in neighbouring bands. When the forehead muscles are trigger the band B11 goes up but at the same time bands B7-B10 also go up, which should not be the case. It was also observed that the bands B1-B3 climb up when the user glances left or right. It only indicates whether the user has glanced in any of the two directions. The electrodes used for EOG should be placed at the extreme end of the forehead to capture signals with higher energy. It is recommended to use more channels for data recording to improve Bio-signal differentiation.

Time Constraint:

During the testing stage it was decide to simulate the wheelchair on a map using MobileSim software. The map was ready but as the algorithm consumed more time in modification. The

54

testing of the navigation commands for the wheelchair were done using a simple map and the simulation of the wheelchair movements was indicated by arrows displayed on screen.

Additional features to the user interface could have been added if more time was available for the interface development and motor speed control part of the code.

5   Conclusion

A controller using hybrid Bio-signals has been developed and tested for an intelligent wheelchair. It was an enriching experience to resolve challenging problems faced throughout the project. Every control system needs to be improvised to reach a level of precision for real world application. Further developments in certain areas have been mentioned to make the system more robust and suitable for individual needs.

My work can be used to assist not only quadriplegic individuals but also paralysed, elderly and individuals with muscle degeneration condition. Using the proposed controller the individual will require minimum human assistance for operating the wheelchair. This project will also give them an opportunity to experience mobility freedom.

6 References

[1]. Axel Lankenau, Thomas Röfer, *Smart Wheelchairs- State of the Art in an Emerging Market,* Zeitschrift Kunstliche Intelligenz 4/00. Schwerpunkt Autonome Mobile Systeme. Fachbereich 1 der Gesellschaft fur Informatik e.V., arenDTaP. 37-39, 2000.

[2]. T. Iberall, G. Sukhatme, D. Beattie, and G. Bekey, *On the Development of EMG Control for a Prosthetic Hand,* Proceeding of IEEE International Conference on Robotics and Automations (ICRA94), 1994.

[3]. A. Barreto, S. Scargle and M. Adjouadi, *A Practical EMG-based Human-Computer Interface for users with motor disabilities,* Journal of Rehabilitation Research and Development vol.37, number 1200.

[4]. J. Gips and P. Oliviere, *Eagle Eyes: An eye control system for persons with disabilities,* Proceedings of IEEE Interenational Conference on Technology and Persons with Disabilities, 1996.

[5]. L.Kirup, A. Searle, A. Craig, P. Mclsaac and P. Moses, *EEG-based System for rapid on off switching without prior learning,* Medical and Biological Engineering and Computing, vol. 35, Pg. 504-509,1997.

[6]. Inhyuk Moon, Myoungioon Lee and Museong Mun, *A Novel EMG-based Human-Computer Interface for persons with Disability,* ICM 2004, Proceeding of IEEE International Conference on Mechatronics, 2004.

[7]. T. Felzer and B. Freisleben, *HaWCoS: The 'Hands-free' Wheelchair Control System,* Proceedings of International ACM SIGACCESS Conference on Computers and Accessibility, pp. 127–134, ACM Press, 2002.

[8]. Pankaj Kadam(2010), "MSc Project Proposal", *School of Computer Science and Electronic Engineering, University of Essex ,* 20 March 2010.

[9]. Herman L. Kamenetz, *A Brief History of the wheelchair,* 1969.

[10]. Ida Bromley, *Tetraplegia and paraplegia: a guide for physiotherapists,* 2006.

[11]. Bell D. A., Borenstein J., Levine S. P., Koren Y., Jaros J., *An assistive navigation system for wheelchairs based upon mobile robot obstacle avoidance,* Proceedings of IEEE International Conference on Robotics and Automations, 1994.

[12]. Roger Bostelman, James Albus, *A multipurpose robotic wheelchair and rehabilitation device for the home ,* International Conference on Intelligent Robots and Systems –IROS, 2007.

[13]. Lai Wei, Huosheng Hu and Kui Yuan, *Use of Forehead Bio-signals for Controlling an Intelligent Wheelchair,* Proceedings of 2008 IEEE International Conference on Robotics and Biomimetics, 2009.

[14]. Sarangi P. Parikh, Valdir Grassi Jr., Vijay Kumar and Jun Okamoto Jr, Incorporating User Inputs *in Motion Planning for a Smart Wheelchair*, International Conference on Robotics and Automations, 2004.

[15]. Bill Sellers, *Introduction to EMG*, University of Manchester, lecture notes, 2002.

[16]. *Brainfingers^{TM} User manual version 6.0*, Brain Actuated Technologies, Inc, 2004.

[17]. N.Nojd and J. Hyttinen, *Modeling of EOG and electrode position optimization for human-computer interface*, ACM proceedings' of ICST 3rd international conference on Body area networks, 2008.

[18]. P. Jia, H. Hu, T. Lu and K. Yuan, *Head Gesture Recognition for Hands-free Control of an Intelligent Wheelchair*, Industrial Robot: An International Journal, 2007.

[19]. Howard Demuth, Mark Beale, Martin Hagan, *Neural Network Toolbox^{TM} User's Guide,*1992-2010 by the Math Works Inc.

[20]. Steffen Nissen, *Neural Networks made simple*, 2005 the FANN library.

[21]. Alexandros Stergiakis, *Tcl fann- a TCL extension for Artificial Neural Network*, 2008.

[22]. M. Riedmiller and H. Braun, *A Direct Adaptive Method for Faster Backpropagation Learning: The RPROP Algorithm*, Proceedings of IEEE International Conference on Neural Networks (ICNN), San Francisco, 1993.

[23]. Richard C. Simpson, *Smart Wheelchairs: A literature review*, Journal of Rehabilitation Research and Development (JRRD), Volume 42, no. 4, pages 423-438, July/August 2005,

[24]. H. Niniss and A. Nadif, *Simulation of the Behaviour of a powered Wheelchair using virtual reality*, Proceedings 3rd International Conference Disability, Virtual Reality and Assoc. Tech., Alghero, Italy, 2000.

[25]. Richard Simpson, Simon P. Levinel, David A. Bell, Yoram Koren, Johann Borenstein and Lincoln A. Jaros, *The NavChair Assistive navigation System*, Presented at the 1995 International Joint Conference on Artificial Intelligence (IJCAI), 1995.

[26]. Taken from *http://www.kinovarehab.com/*, on 25th August 2010.

[27]. Taken from *http://odalab.spub.chitose.ac.jp/english/research.html* , on 25th August 2010.

[28]. Richard Simpson, Edmund LoPresti, Steve Hayashi, Songfeng Guo, Dan Ding, William Ammer, Vinod Sharma, Rory Copper, *A Prototype Power assist wheelchair that provides for obstacle detection and avoidance for those with visual impairments,*Journal of NeuroEngineering and Rehabilitation, *2005.*

[29]. Taken from *http://medgadget.com/archives/2006/10/hlpr_chair.html,* on 28th August 2010.

[30]. Taken from *http:// www.autoblog.com/photos/toyota-bmi-wheelchair/#2114222,* on *28th August 2010.*

[31]. Taken from *http://www.nia.nih.gov/Alzheimers/publications//UnraveelingtheMystery/, on 29th July 2010.*

[32]. Chun Sing Louis Tsui, Pei Jia, John Q. Gan, Huosheng Hu and Kui Yuan, EMG-based Hands-Free Wheelchair Control with EOG Attention Shift Detection, Proceedings of IEEE International Conference on Robotics and Biomimetics, China, 2007.

About the author

Pankaj Raghunath Kadam was born in Mumbai, India in 1988. He completed his Masters in Embedded Systems at the University of Essex, England in 2010. During his studies and after, he worked as an AMX programmer at the ISS (Information System Services) department at the University.

The author has received the Educational Achievement award for his Master's degree and Award of Excellence during his Bachelor studies in Engineering.

Currently, the author is back in Mumbai and works as an Embedded Scientist at Jhaveri Labs. When he isn't glued to the computer, he spends time playing guitar and enjoys different aspects of artwork.

www.ingramcontent.com/pod-product-compliance
Lightning Source LLC
LaVergne TN
LVHW042350060326
832902LV00006B/506